active minds® EXPLORERS

Baby PANDA

Ellen Lawrence

sequoia™ kids media

Photography © Shutterstock 2021 9dream studio; Alex Kednert; Anastasiya Zaplatina; astudio; avkost; BrAt82; bsd; bychovsky; DistanceO; DSBfoto; Flas100; gladcov; Greenni; hchjjl; Inka1; Irmun; Juhku; Kamieshkova; kovalto1; lalan; Macrovector; Max Krasnov; Milanares; Nastyuha creative; Nik Merkulov; Olex Runda; optimarc; Pyty; sociologas; th3fisa; vasabii; Veronika Yeroshenko; VolodymyrSanych; xpixel; Yaska; zhaoliang70, Superstock

Published by Sequoia Kids Media,
an imprint of Sequoia Publishing & Media, LLC

Sequoia Publishing & Media, LLC.,
a division of Phoenix International Publications, Inc.

8501 West Higgins Road
Chicago, Illinois 60631

Customer Service: cs@SequoiaKidsBooks.com

www.SequoiaKidsMedia.com

ISBN 978-1-64996-219-5

Baby PANDA

TABLE OF CONTENTS

Words shown in **bold** in the text are explained in the glossary.

A BABY IN A DEN

In a **bamboo** forest, there is a small cave.

The cave is the cozy **den**, or home, of a tiny baby.

The little animal cannot see or walk.

She is so small, she
could fit into a person's hand.

Who does this little baby belong to?

MOTHER GIANT PANDA
PAW
A two-day-old panda cub

The tiny baby, or cub,
belongs to a mother giant panda.

The mother panda gently holds
the cub in her mouth or paw.

The cub drinks milk from her mother.

Safe inside the den, the little panda cub grows bigger.

By the time she is three weeks old, she has black legs, ears, and eye patches—just like her mom!

A three-week-old panda cub

The mother panda keeps the little cub warm with her big, furry body.

She does not leave the den until the cub is four weeks old.

A four-week-old panda cub

During this time, the mother panda has nothing to eat or drink.

By the time the cub is six weeks old, she has thick fur and her eyes have opened.

The mother panda licks the cub to keep her clean.

A two-month-old panda cub

A three-month-old panda cub

When the mother panda goes to find food, the cub waits in the den.

When the cub is four months old, she can walk and run.

She plays with her mom outside the den.

When playtime is over, she enjoys a drink of milk.

Pandas are very good at climbing trees.

When the cub is five months old, she climbs...

...and climbs...

...and climbs.

Sometimes, she sits in a tree on her own for several hours!

A panda's main food is bamboo.

MOTHER PANDA

When the cub is about six months old, she tries this tough, crunchy food.

She still drinks her mother's milk, though.

A six-month-old panda cub

When the cub is two years old, she is ready to leave her mom.

Now, the young panda lives alone in the bamboo forest.

One day, she will be ready to have a cub of her own.

FACT FILE

All About Pandas

Adult male giant panda

Giant pandas are a type of bear.

A female panda has her first cub when she is about six years old.

Adult pandas live alone. Males and females meet up when it's time to **mate**.

Each panda lives in its own area called a **territory**. It marks the edges of its territory by clawing tree trunks and spraying them with pee.

Father pandas do not help take care of their cubs.

Panda Size

Panda Weight

Adult Panda:
220 to 330 pounds (100–150 kg)

Newborn Panda:
3.5 ounces (100 g)

A newborn panda is about 7 inches (18 cm) long.

Where Do Pandas Live?

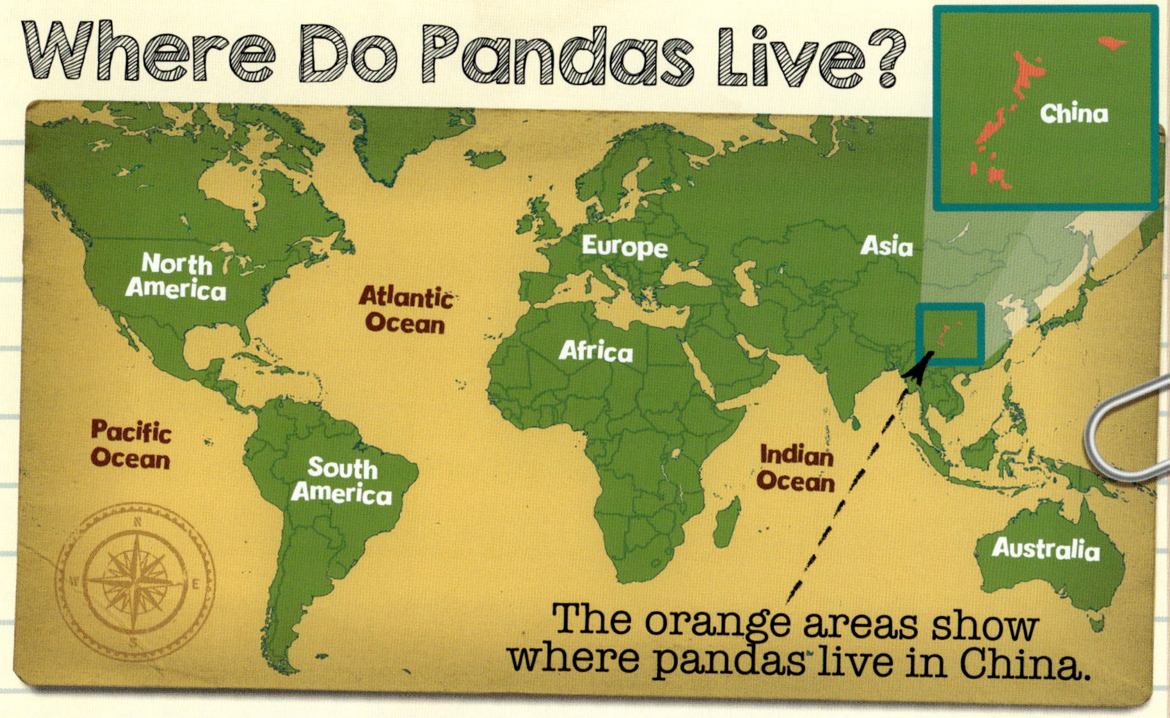

Pandas in Danger

There are only about 1,800 pandas living wild in China.

The bamboo forests where they live have been cut down by people to make space for roads and towns.

Pandas need the forests as a place to live and find food.

The government of China and organizations such as WWF (World Widelife Fund for Nature) are turning areas of forest into special protected parks.

Inside the parks, people are not allowed to cut down the forests.

Glossary

bamboo (bam-BOO)
A type of fast-growing grass plant with thick, hollow stems.

den (DEN)
An animal's home. A den might be inside a cave or tree stump. It might also be underground.

mate (MATE)
To get together to have babies.

territory (TER-uh-tor-ee)
An area where an animal lives and finds its food. An animal might try to defend its territory from rivals to protect its food.

Index

Read More

From Cub to Panda
(Start to Finish)
Jennifer Boothroyd
Minneapolis, MN: Lerner Publishing
2017

Panda
(Wild World)
Meredith Costain
New York, NY: Windmill
2017

Visit Us

www.SequoiaKidsMedia.com

Downloadable content and more!

N
E
S